Learning About the Changing Seasons

Dear Parents,

Young children are natural scientists, curious about the world around them. They have an infinite capacity to learn and are eager to know why and how things work the way they do. *Little Scientists, Hands-On Activities* begins with the simple questions most children ask and then shows them how to explore and discover for themselves. Our acclaimed Little Scientists, "hands-on" approach instills in children a passion for the exciting world of science and helps children develop specific scientific skills that will provide a strong foundation for later learning.

With this book, you can join me on a journey into the wonders of the changes that take place on Earth. Together we will discover the unlocked secrets of spring, summer, fall, and winter, and learn how to create a sun oven, a snow catcher, and many other exciting projects.

Your Little Scientist can e-mail me at

Dr_Heidi@Little-Scientists.com

Wishing you many enjoyable discoveries,

Dr. Heidi

Dr. Heidi

Little Scientists®

A "hands-on" approach to learning

Learning About the Changing Seasons

Heidi Gold-Dworkin, Ph.D.

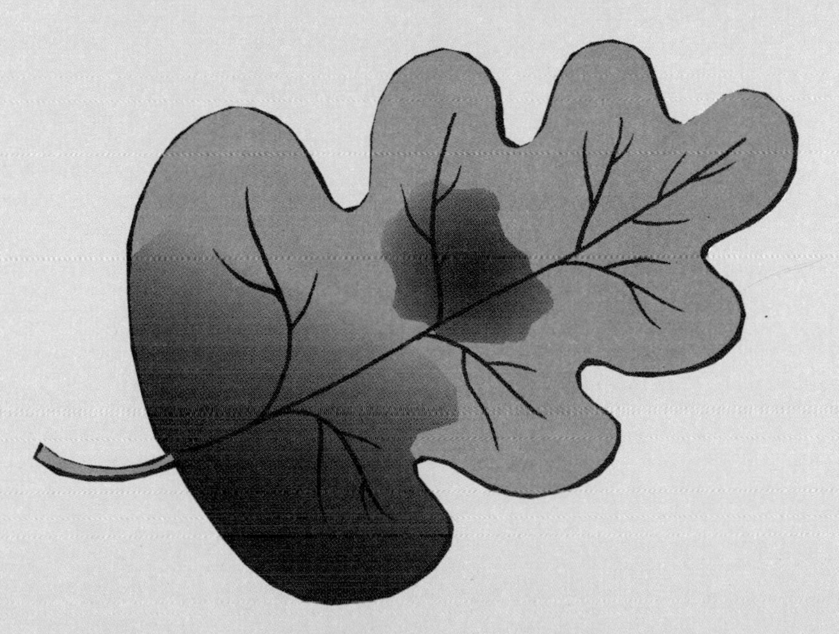

McGraw-Hill

New York San Francisco Washington, D.C.
Auckland Bogotá Caracas Lisbon London Madrid Mexico City
Milan Montreal New Delhi San Juan Singapore Sydney Tokyo Toronto

This book is dedicated to my children
Aviva, Olivia, and Robert

This book would not have been possible
without the contributions from the following
staff members at Little Scientists:®

Ronda Margolis

Avi Ornstein

June Stevens

Linda Burian

Bec Luty

Donna Goodman

I would like to thank my devoted family, especially
my husband, Jay; mom, Jacqueline; and sister, Stacey.

McGraw-Hill
A Division of The ***McGraw-Hill*** *Companies*

pbk 3 4 5 6 7 8 9 0 QPD / QPD 0 9 8 7 6 5 4 3 2 1

ISBN 0-07-134822-0

Library of Congress Cataloging-in-Publication data applied for.

McGraw-Hill books are available at special quantity discounts to use as premiums
and sales promotions. For more information, please write to the Director of Special Sales,
McGraw-Hill, Two Penn Plaza, New York, NY 10121-2298. Or contact your local bookstore.

Acquisitions editor: Mary Loebig Giles
Senior editing supervisor: Patricia V. Amoroso
Senior production supervisors: Clare B. Stanley and Charles Annis
Left page illustrations: Robert K. Ullman <r.k.ullman@worldnet.att.net>
Right page illustrations: K. Almadingen <dzbersin@aol.com>
Book design: Jaclyn J. Boone <bookdesign@rcn.com>

Printed and bound by Quebecor/Dubuque.

This book is printed on recycled, acid-free paper containing
a minimum of 50% recycled, de-inked fiber.

Contents

Spring
Summer

Fall
Winter
Hi, I am Dr. Heidi.
My Little Scientists® friend, Olivia,
is going to breeze through the seasons
with us. We are going to discover
changes on the Earth that happen
during spring, summer,
fall, and winter.

It's my birthday!
Why can't it be my birthday every day?

There are 365 days in one year and each day is different.
This experiment will show you why your birthday only comes once a year.

1. Take the calendar and count the days with the help of an adult. Do you count 365 days?

You will need
- A calendar

2. How many months are there?

January	July
February	August
March	September
April	October
May	November
June	December

Do you count 12 months?

JANUARY

						1
2	3	4	5	6	7	8
9	10	11	12	13	14	15
16	17	18	19	20	21	22
23	24	25	26	27	28	29
30	31					

FEBRUARY

		1	2	3	4	5
6	7	8	9	10	11	12
13	14	15	16	17	18	19
20	21	22	23	24	25	26
27	28	29				

MARCH

			1	2	3	4
5	6	7	8	9	10	11
12	13	14	15	16	17	18
19	20	21	22	23	24	25
26	27	28	29	30	31	

3. Olivia's birthday is on March 21. How many times do you find March 21 on the calendar?

MY BIRTHDAY

APRIL

						1
2	3	4	5	6	7	8
9	10	11	12	13	14	15
16	17	18	19	20	21	22
23	24	25	26	27	28	29
30						

MAY

	1	2	3	4	5	6
7	8	9	10	11	12	13
14	15	16	17	18	19	20
21	22	23	24	25	26	27
28	29	30	31			

JUNE

				1	2	3
4	5	6	7	8	9	10
11	12	13	14	15	16	17
18	19	20	21	22	23	24
25	26	27	28	29	30	

Each day comes only once a year!

JULY

						1
2	3	4	5	6	7	8
9	10	11	12	13	14	15
16	17	18	19	20	21	22
23	24	25	26	27	28	29
30	31					

AUGUST

		1	2	3	4	5
6	7	8	9	10	11	12
13	14	15	16	17	18	19
20	21	22	23	24	25	26
27	28	29	30	31		

SEPTEMBER

					1	2
3	4	5	6	7	8	9
10	11	12	13	14	15	16
17	18	19	20	21	22	23
24	25	26	27	28	29	30

OCTOBER

1	2	3	4	5	6	7
8	9	10	11	12	13	14
15	16	17	18	19	20	21
22	23	24	25	26	27	28
29	30	31				

NOVEMBER

			1	2	3	4
5	6	7	8	9	10	11
12	13	14	15	16	17	18
19	20	21	22	23	24	25
26	27	28	29	30		

DECEMBER

					1	2
3	4	5	6	7	8	9
10	11	12	13	14	15	16
17	18	19	20	21	22	23
24	25	26	27	28	29	30
31						

The birds are back. It must be spring!
Where have the birds been all winter?

Every day, you wake up in the morning after sleeping through the night. Every year, much of nature sleeps through the winter and wakes up in the spring! As the days get warmer, the sun shines longer, trees know that it is time to open up their leaves, and the birds that flew away last fall know it is time to return. Let's do an experiment to find out why the world wakes up in the spring.

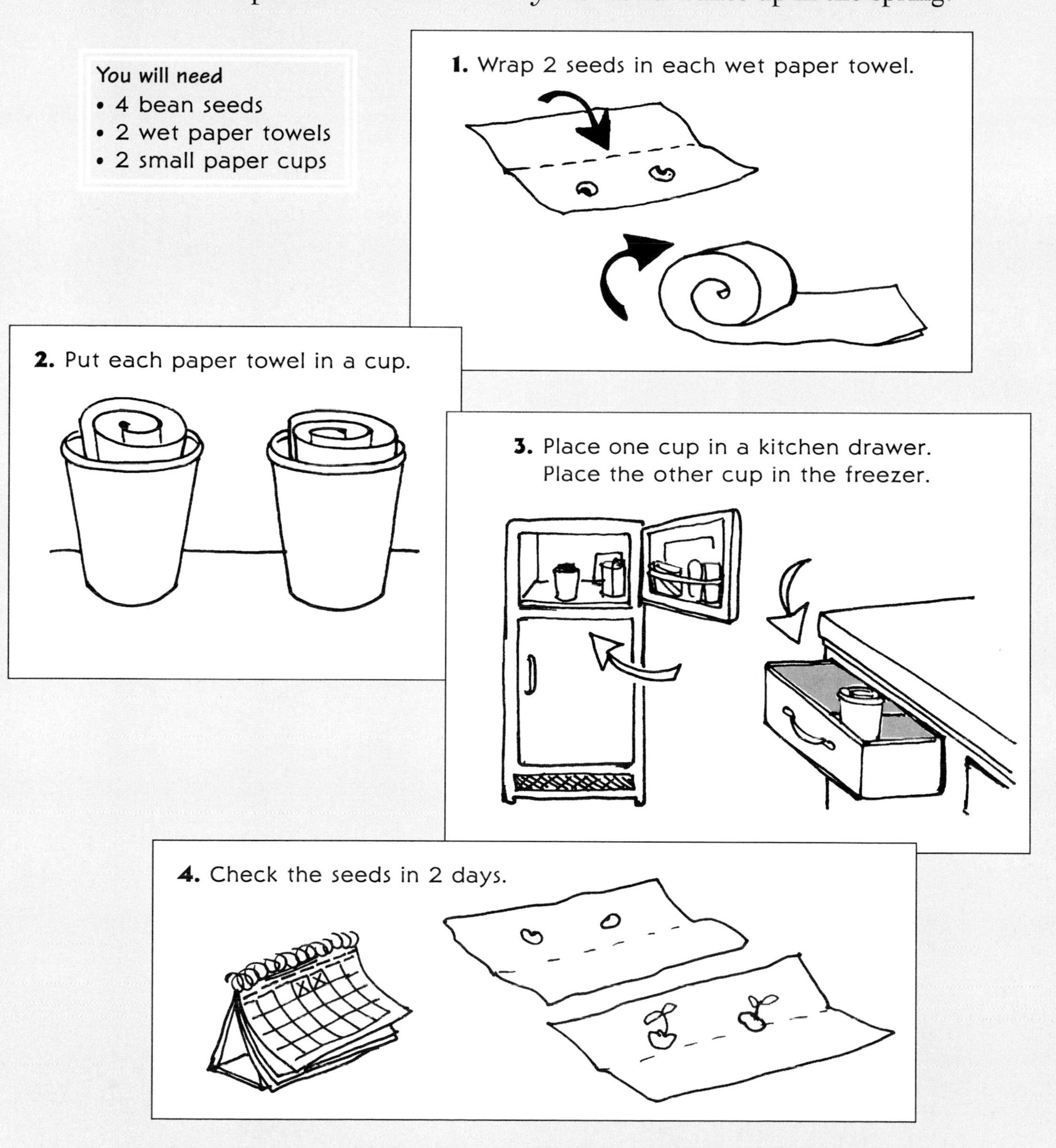

The seeds in the freezer will not grow, but the seeds that were in the drawer will. This experiment shows that living things need warmth to grow.

Do plants need light to grow?

In the previous experiment we learned that living things need warmth to grow. Let's do another experiment to find out if plants also need light to grow.

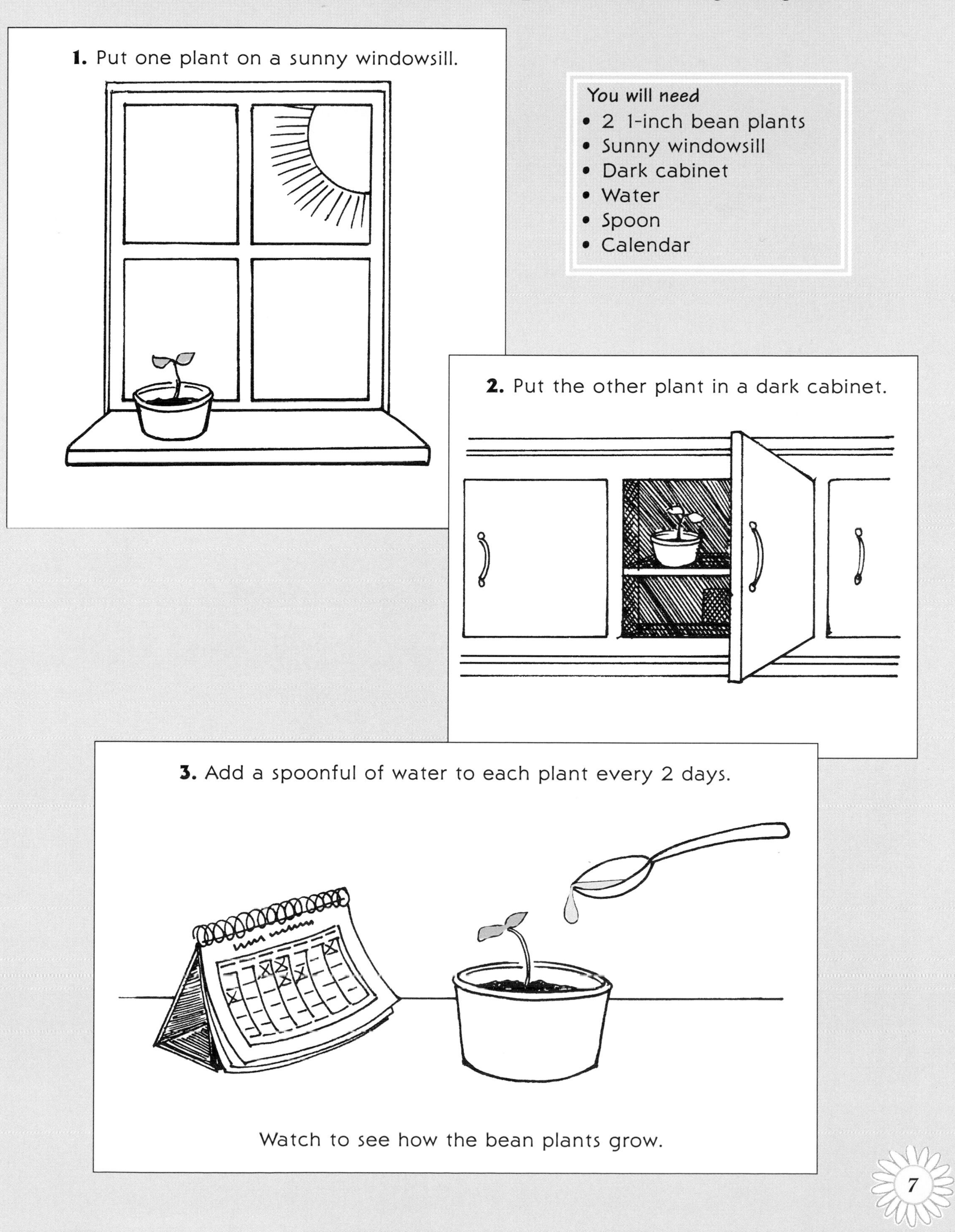

How will grass grow from these seeds?

The process of seeds changing into plants is called **germination**. It requires water and warmth from the sun. Let's do an experiment to see grass seeds change into grass plants.

You will need
- 12" piece of string
- Clear plastic bottle with a wide opening
- Open pine cone that will fit in the top of the bottle
- Dirt
- Water
- Spoon or craft stick
- Paper plate
- 1 spoonful of grass seed
- Sunny window

1. Tie the string to the stem of the pine cone.

2. Mix dirt with a small amount of water until the dirt becomes damp.

3. Using the spoon, spread the dirt into the scales of the pine cone.

4. Put a spoonful of grass seed on a paper plate. Roll the pine cone on the plate until the pine cone is covered with grass seed.

5. Add water to the bottle until the bottom is covered.

6. Tie the pine cone string to the middle of the spoon or craft stick. Place it over the mouth of the bottle so that the pine cone hangs inside.

7. Place the bottle in a sunny window. Observe the bottle every day. How many days do you think it will take the grass to start growing?

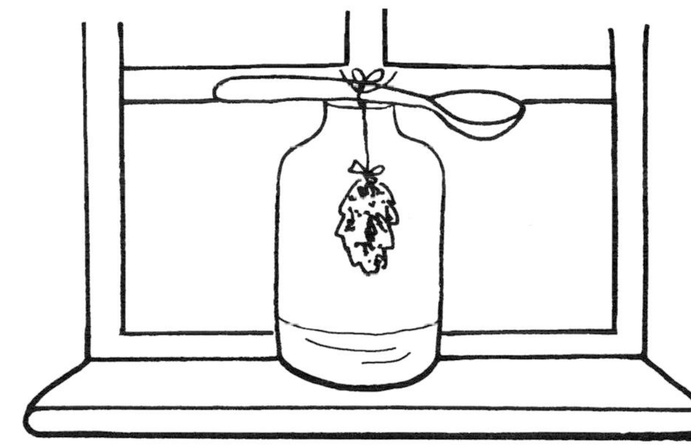

What happens to the plants
under the ground?

Seeds are really little plants that have not yet started to grow. When a seed absorbs water (either from rain or from being added to the garden), it gets fatter and fatter. As the seed keeps growing, its protective skin pops off and the roots begin to grow. The roots grow down into the soil and the seed is pushed up toward the surface. Leaves form as it grows toward the sunlight. You can plant a seed and watch it grow.

What kind of soil do plants grow best in?

This next experiment will help you decide which soil is better for plant growth: clay or dirt.

The sunlight and water are the same for each cup, so it is only the type of soil that is varied. The plants that have the best soil will grow the best!

What is in an egg?

An egg is the beginning of an animal, just like a seed is the beginning of a plant. Under the right conditions, an egg will become an animal! Let's look at what is inside an egg.

1. Look at the shells of the eggs. This is what protects what is inside the egg.

You will need
- 1 raw egg
- 1 hard-boiled egg
- Bowl
- Plate
- Assistance from an adult
- Butter knife

2. Hold the raw egg in one hand. Carefully crack the shell in half on the edge of the bowl.

Using both hands, slowly separate the two halves so that the inside of the egg goes into the bowl. Put the shell aside.

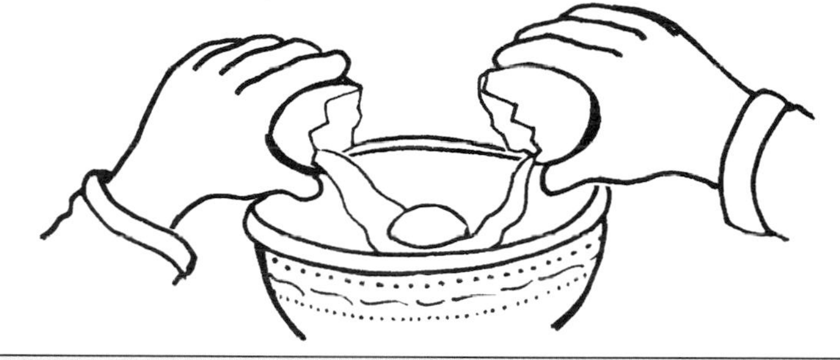

3. Crack and peel the shell off the hard-boiled egg and place it on the plate.

4. Have an adult slice the hard-boiled egg in half.

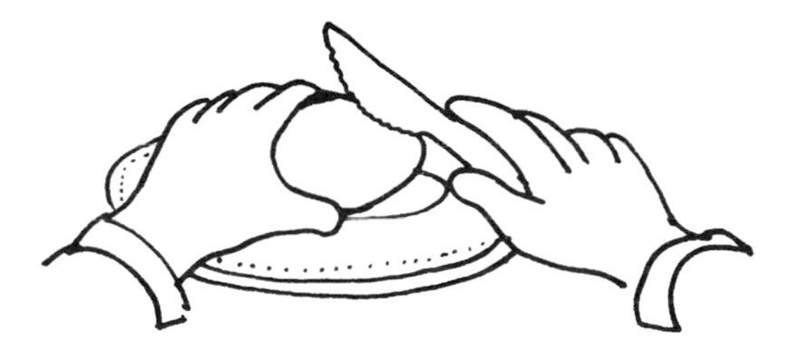

Look at the different parts of both eggs. Why do you think there are two parts?

The yellow part of the egg (the **yolk**) could have become a bird. The white, or clear part (the **albumen**) would have been its food as it was growing inside the shell.

Why is it hotter in the sun than in the shade?

Sunlight brings both heat and light from the sun to the Earth. We get more sunlight in the summer, which is why it is hotter than the other seasons. Here is an experiment that proves that heat comes with the sun's light.

*You will **need***
- Assistance from an adult
- 2 metal cans
- Black enamel paint
- Disposable paint brush
- Sunny spot outside
- Shady spot outside
- Timer or clock

1. Have an adult help you paint the outsides of both cans.

2. When the paint is dry, place one can outside in a sunny spot.

Place the other can in a shady spot.

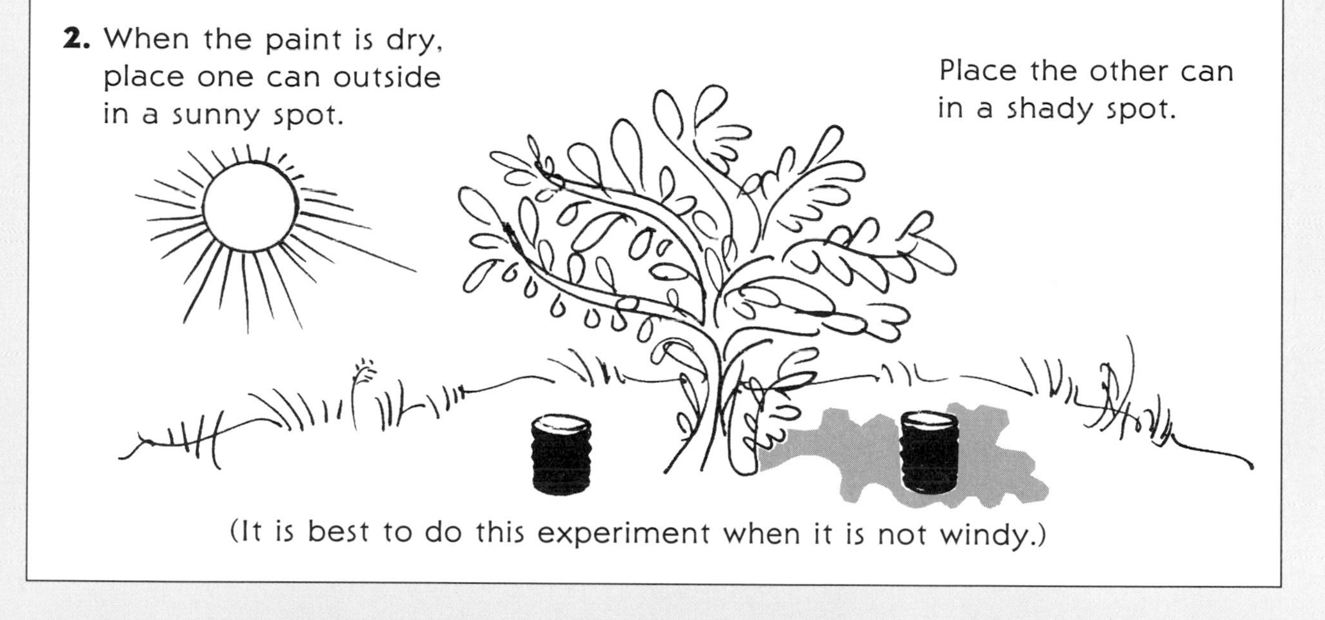

(It is best to do this experiment when it is not windy.)

3. After 15 minutes, go and feel the temperature of each can with your hand.

The black can sitting in the sunny spot has been absorbing energy from the sunlight. That is why it is warmer.

Does anything grow or live under the water?

A pond is a good place to learn about nature. If you look closely, you might see fish, crayfish, tadpoles, and aquatic insects swimming in the water. There are many things in the pond you can't see with your eyes. If you use a small microscope or magnifying glass, you might see very tiny plants and animals that live in a pond. Let's make a simple microscope and try it out!

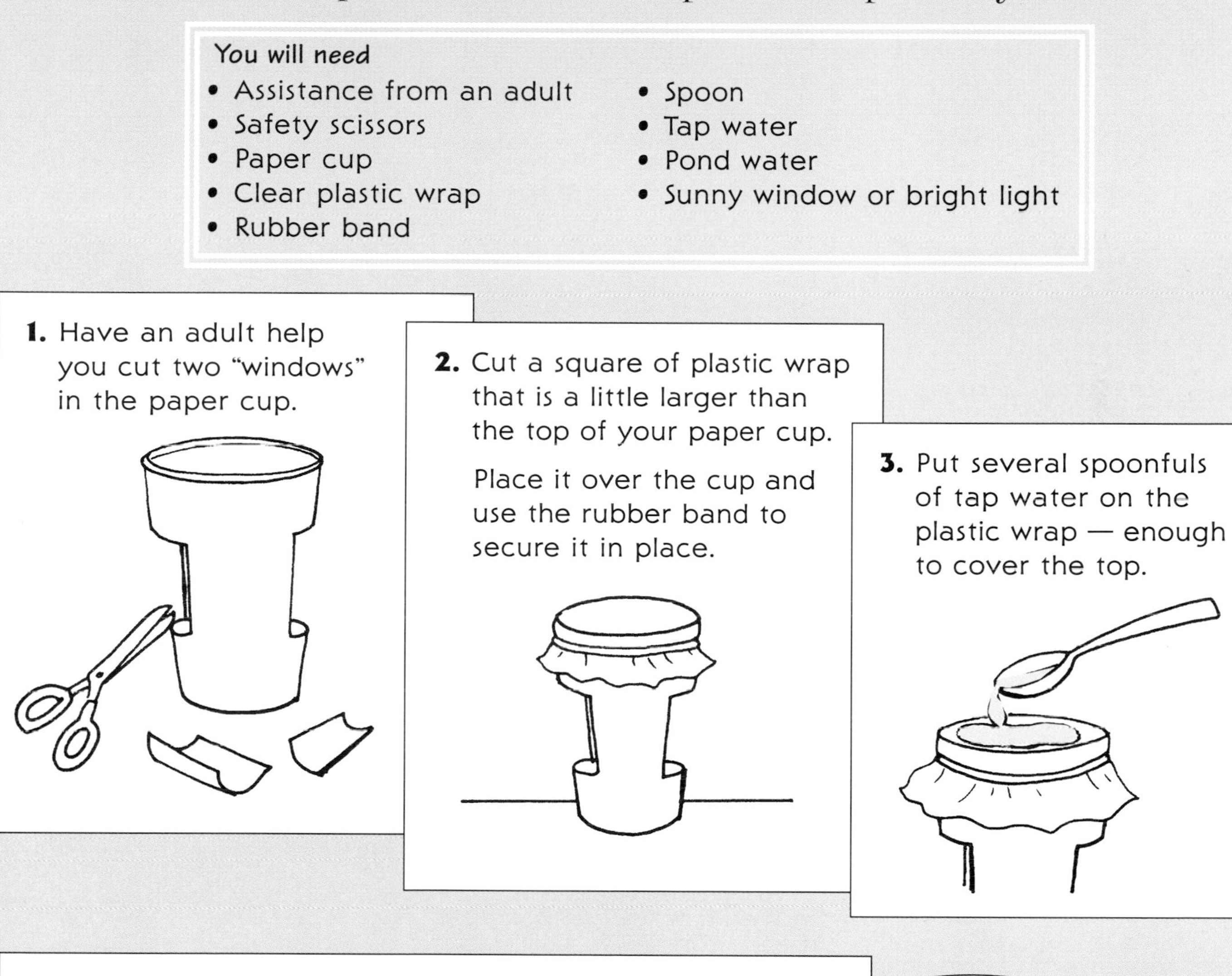

You will need

- Assistance from an adult
- Safety scissors
- Paper cup
- Clear plastic wrap
- Rubber band
- Spoon
- Tap water
- Pond water
- Sunny window or bright light

1. Have an adult help you cut two "windows" in the paper cup.

2. Cut a square of plastic wrap that is a little larger than the top of your paper cup.

Place it over the cup and use the rubber band to secure it in place.

3. Put several spoonfuls of tap water on the plastic wrap — enough to cover the top.

4. Place a sample of several spoonfuls of the pond water in the bottom of the cup. Place the cup near a sunny window or bright light.

Look through the top of the cup to see the enlarged image of what is in the pond water.

Why is mud at the bottom of the pond?

In still water, such as a pond, stirred-up sand and dirt settle to the bottom with a layer of mud on top. Let's do an experiment to test this out.

Heavier particles settle faster, so the sand quickly falls to the bottom of the jar. The dirt is made up of very small particles. It will take a while for all of them to settle on the bottom. The water gets cleaner as the dirt settles down and a layer of mud will form on top of the sand. The same thing happens in a pond.

What made my candy bar melt?

The energy from the sun heats the Earth. It is hottest when our portion of the Earth is most directly facing the sun. This happens during the summer and during the middle of the day. That is why it is hottest in the middle of summer days. Let's see how you can collect and use this heat.

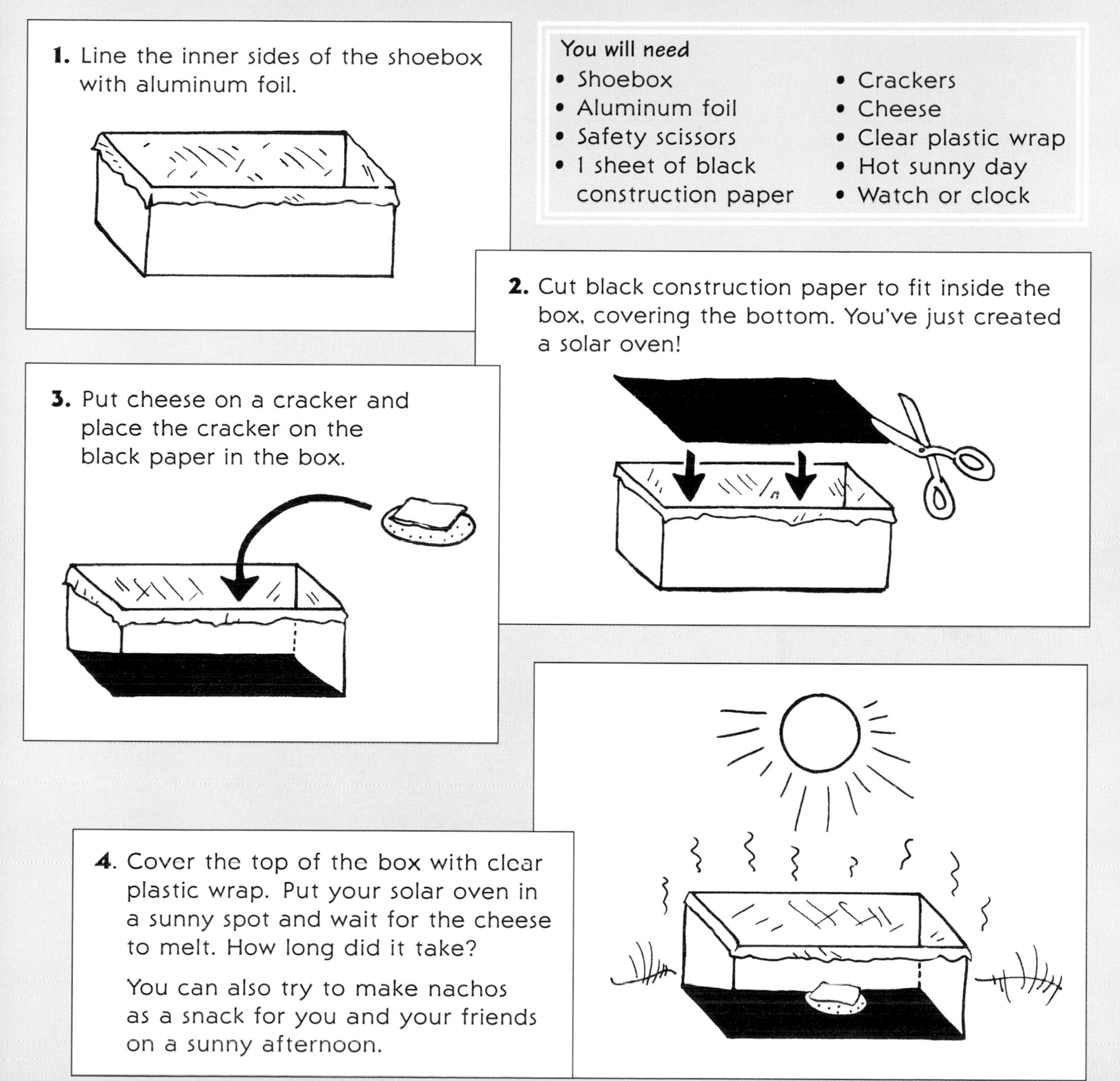

The black construction paper caught and absorbed the heat from the sun. The aluminum foil reflected more sunshine down on the bottom of the box. The paper released the heat to the air above it, making the air hotter. The plastic wrap trapped the heated air, so it got hotter and hotter inside the box! This is the way solar panels trap heat. The same process explains why a car with closed windows gets hot during the summer.

What's inside a flower?

Flowers come in all different colors. They are important to plants, because they make seeds so that new plants can grow. If you study flowers, you can see that they are made of petals that surround smaller parts. This next activity helps you see what is inside a flower.

You will need

- Flowers from your garden or from a florist (that have 5 or 6, petals such as lily, petunia, gladiola)
- Plastic knife
- 8 1/2" x 11" sheet of construction paper
- 8 1/2" x 11" piece of wax paper
- Heavy book
- 8 1/2" x 11" piece of clear contact paper
- Marker

1. Carefully cut the flower in half lengthwise, using the plastic knife.

2. Lay the opened flower on the construction paper.

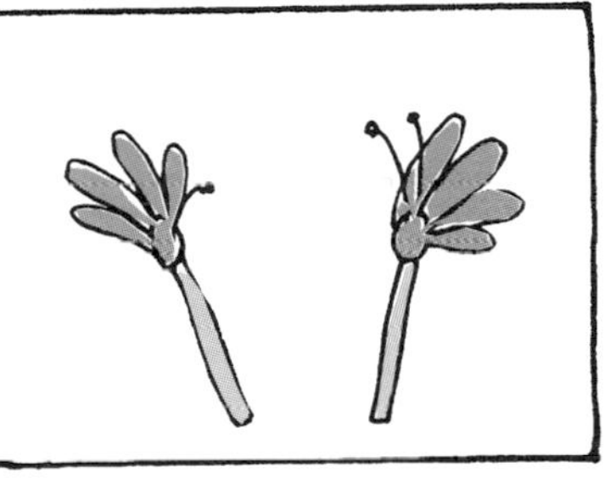

3. Place the wax paper on top of the flower.

4. Place the heavy book on top of the wax paper. Let it remain untouched for one day.

5. On the next day, remove the book. Then carefully remove the wax paper.

6. Peel the back off of the clear contact paper. With the sticky side down, press it on top of the dissected flower to laminate it onto the construction paper.

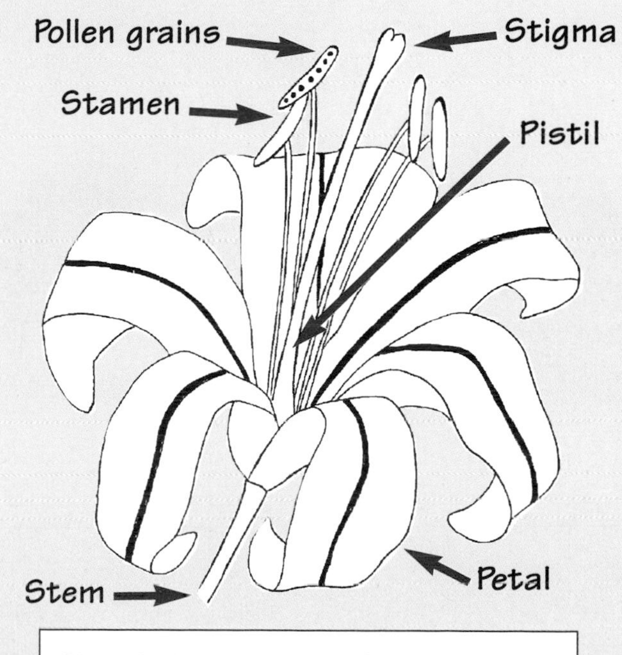

7. Label the parts of the flower, using this picture as a guide.

Why do bees visit flowers?

The colorful petals and the sweet smell of the flower attracts bees (and other animals, such as birds and butterflies).
The bees collect nectar to make honey. **Pollen** sticks to their bodies and is transferred from flower to flower. Plants use the pollen to make seeds.
Let's do an experiment to understand how the pollen is transferred.

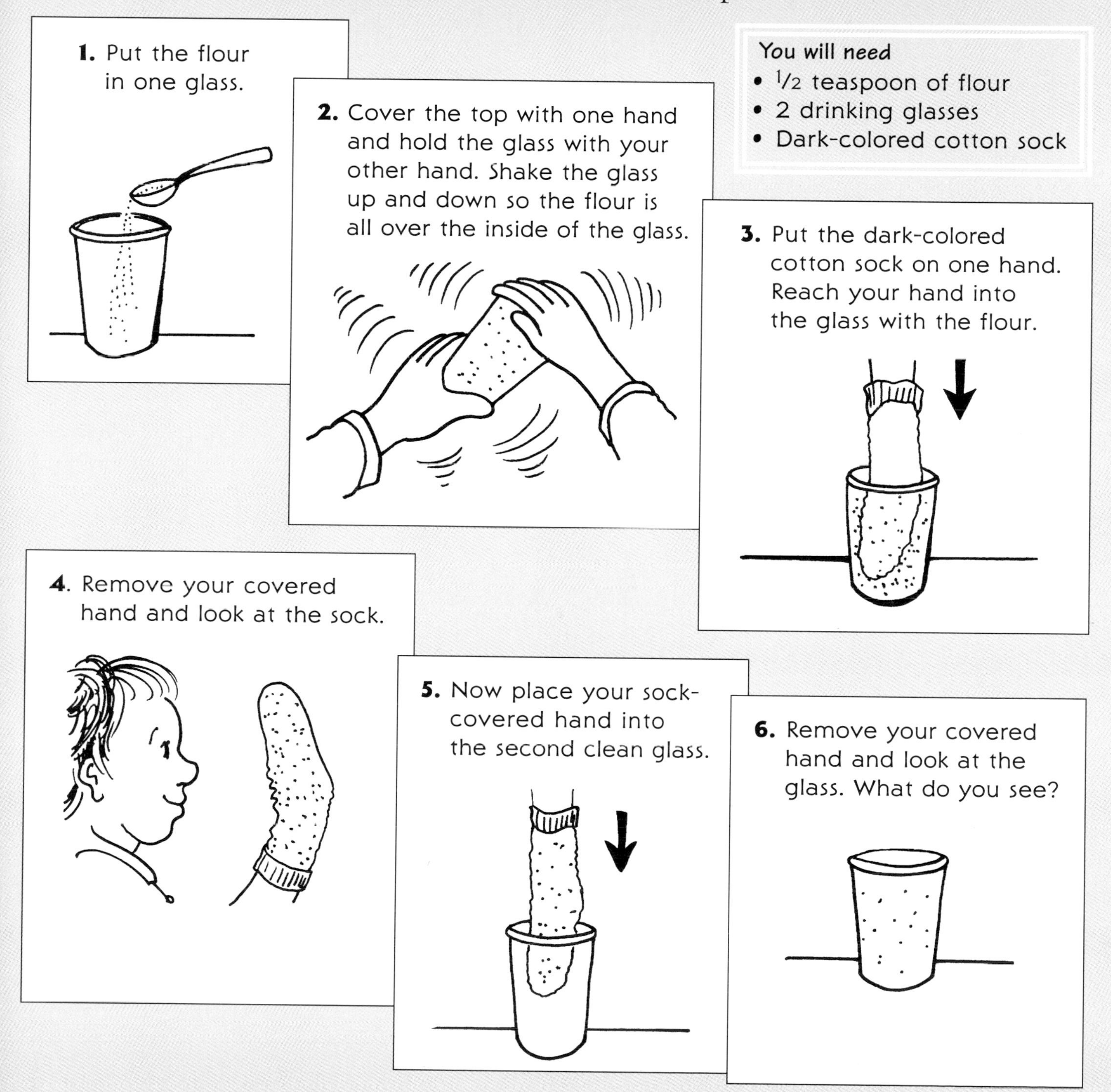

Pretend that the glasses were two flowers. The flour was pollen and your covered hand was a bee. You went into the first flower to collect **nectar** (the flour) to make honey. When you came out of the flower, pollen was stuck to your body. When you went into the second flower, you left some pollen behind. This is what happens when the bees go from one flower to the next.

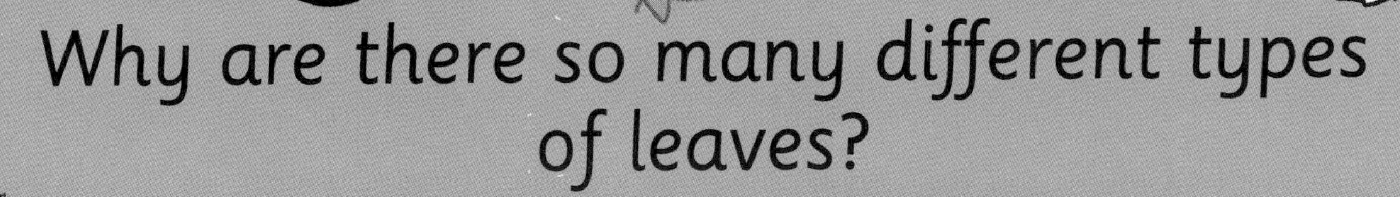

Why are there so many different types of leaves?

ADULT HELP REQUIRED

Different trees have different leaves. Some trees, such as the sassafras, have more than one type of leaf on the same tree! Let's make a preserved leaf book.

You will need

- Leaves of different shapes, sizes, and colors that you find on the ground
- 8 to 10 sheets of 8 1/2" x 11" wax paper
- Newspaper
- Steam iron
- Glue
- 1" wide strips of white or yellow construction paper
- Pencil or marker
- Hole punch
- String or yarn

1. Place a leaf between two pieces of wax paper.
2. Then place a sheet of newspaper over the wax paper.

Newspaper
Wax paper
Leaf
Wax paper

3. Have an adult press gently across the top of the newspaper with a warm steam iron. The pieces of wax paper will melt together and will seal the leaf between them.

   Remove the newspaper.

4. Glue construction-paper strips along the edges of the wax paper, creating a frame. Write the name of the leaf on the frame border.

(You can get a book at the library to help identify leaves, or you can search the Internet to learn more about trees.)

5. Repeat steps 1 – 4 with other leaves.
6. Punch two holes on the left side of each leaf page and connect them, using string or yarn.

Why are there different kinds of seeds?

Just as there are many different types of leaves, there are many different types of seeds. Seeds have different means of **dispersal** — getting to places where new plants can grow. You can learn a lot about plants by collecting their seeds!

You will need

- Seeds and nuts from your yard (or from a hike in the woods with your family)
- Egg carton
- 8 1/2" x 11" sheet of paper
- Safety scissors
- Pencil or marker
- Tape

1. Put each seed or nut in a separate section of the egg carton.

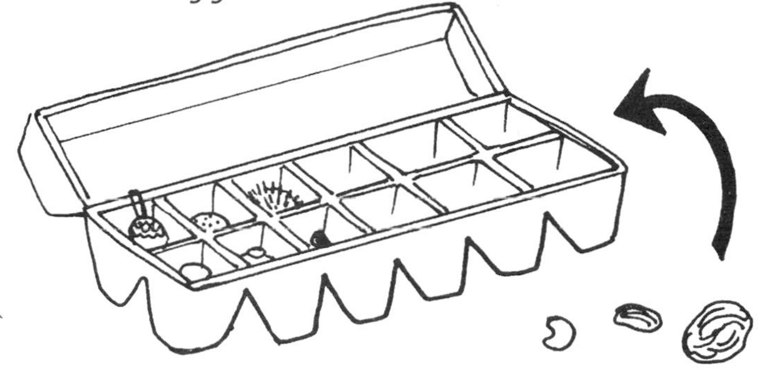

2. Identify the seeds and nuts you have collected.

(You can get a good book at the library to help identify nuts and seeds, or you can try searching the Internet to find more information about seeds.)

3. Make labels for the names of the seeds and nuts, writing each name on the paper. Then cut them out.

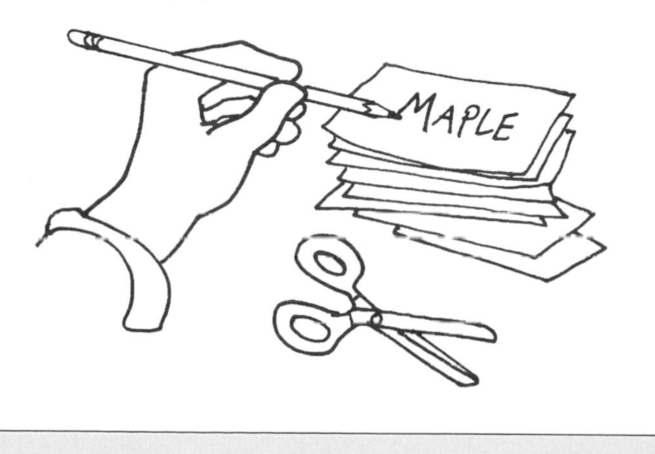

4. Put each label in the appropriate egg carton section and tape it in place.

Why does an apple have a peel?

Apples have peels to protect the seeds inside.
Let's do an experiment to see what happens to an apple that doesn't have a peel.

Without the protection of the peel, the apple will lose water and shrink.
The lemon juice and salt will bleach the skin color of the apple head.

Where do the colors of leaves come from?

Throughout spring and summer, leaves are green. The other colors we see in the fall — red, yellow, and orange — were already present in the leaves. We can discover these colors by using a scientific process called **chromatography**.

You will need

- Fresh spinach leaves
- Porcelain or stoneware coffee mug
- Pinch of sand
- Smooth, round rock
- Spoon
- Fingernail polish remover
- Safety scissors
- Round coffee filter
- Toothpicks
- Measuring cup
- Rubbing (isopropyl) alcohol
- Paper cup
- Tape
- Construction paper
- Ruler

1. Break up the spinach leaves and put the pieces into the mug.

2. Add a pinch of sand.

3. Using the smooth, round rock, grind and squish the leaves into smaller pieces.

4. Add a spoonful of fingernail polish remover to extract the pigments from the leaves.

5. Keep grinding the leaves with the stone until you see the liquid has gained some color from the spinach.

6. Use the safety scissors to cut the coffee filter into a 2" x 8" rectangle.

Continued on the next page.

7. Use a toothpick to pick up a drop of the colored liquid from the leaf grinding. Place the drop one inch from the end of the coffee filter rectangle. Let it dry.

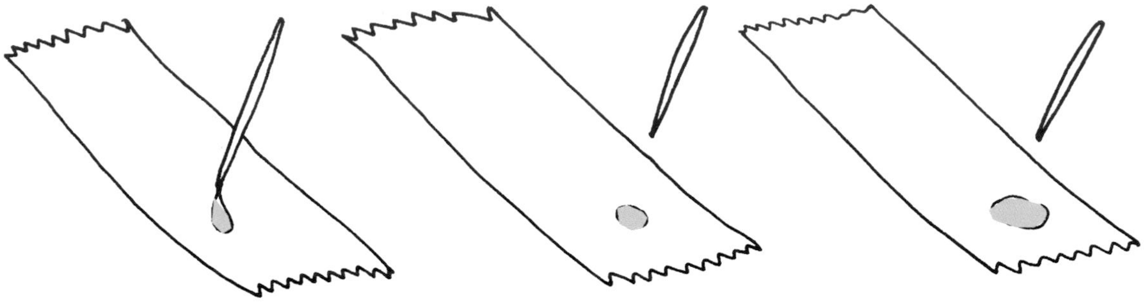

Then use the toothpick again to add a few more drops to the same spot. Let each drop dry before you add the next drop.

8. Put 1/4 cup of alcohol in the paper cup.

RUBBING ALCOHOL

1/4

9. Carefully put the end of the filter paper strip — the end with the drops of dried spinach extract on it — into the alcohol, but do not let the colored drop touch the alcohol. The alcohol will travel up the filter paper, separating the different colors in the spinach. This could take up to an hour.

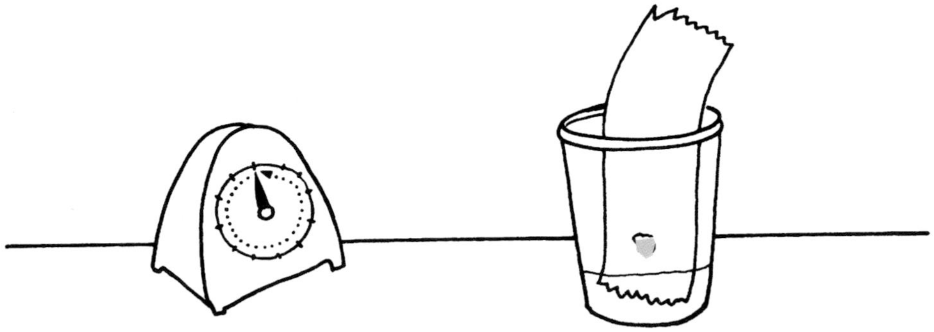

Continued on the next page.

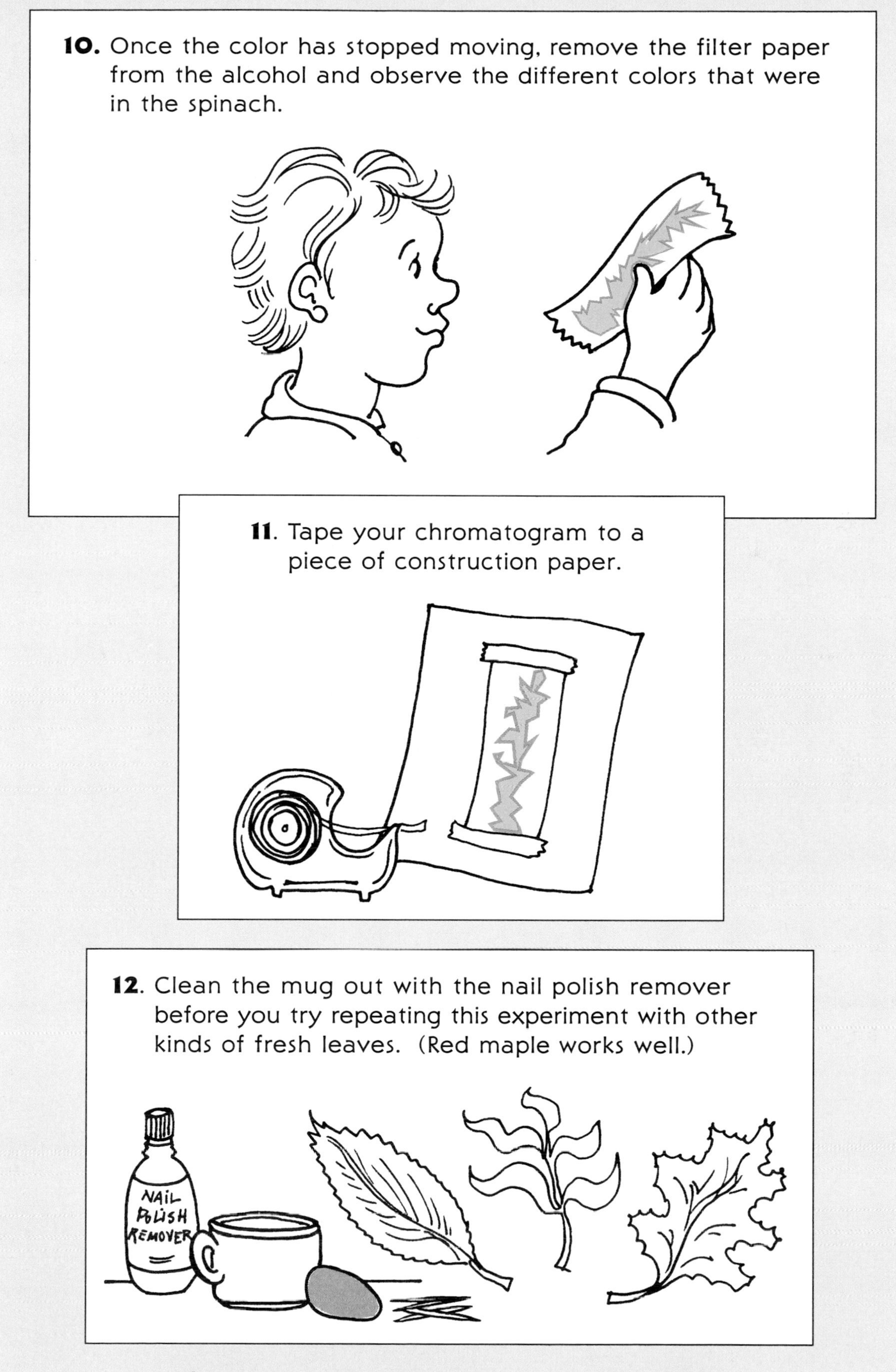

10. Once the color has stopped moving, remove the filter paper from the alcohol and observe the different colors that were in the spinach.

11. Tape your chromatogram to a piece of construction paper.

12. Clean the mug out with the nail polish remover before you try repeating this experiment with other kinds of fresh leaves. (Red maple works well.)

This experiment showed a way to see all the different colors that are present in leaves.

But why do leaves change color in the first place?

Plants lose water **vapor** through their leaves.

Plants have trouble getting water from the ground in the winter because the ground is frozen. In order for plants to keep what water they do have and survive the winter, they cannot lose any water through their leaves. When the leaves stop receiving water, they change color and die, but the rest of the plant survives.

Let's do an experiment to see what happens when the ground is frozen.

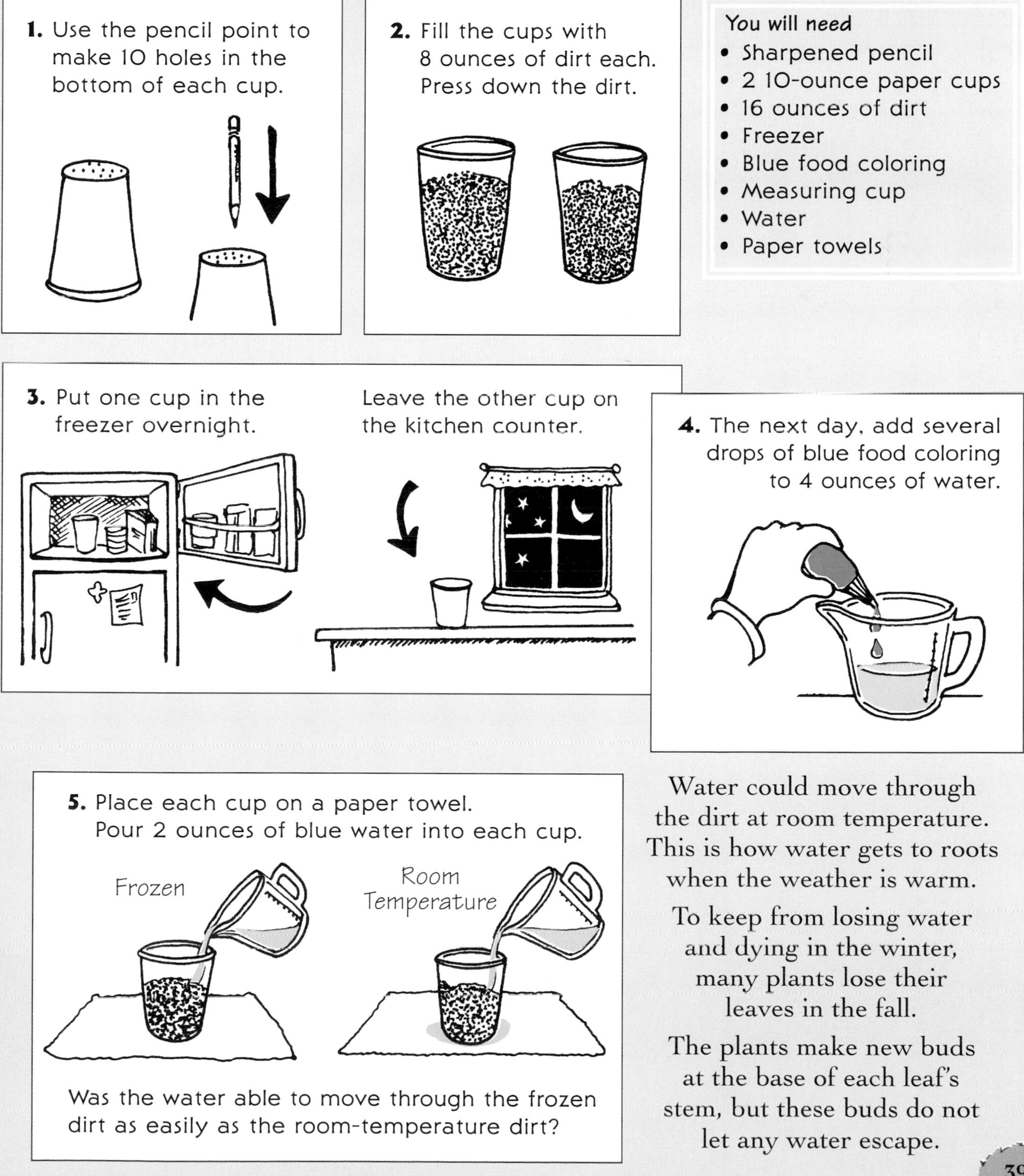

Water could move through the dirt at room temperature. This is how water gets to roots when the weather is warm.

To keep from losing water and dying in the winter, many plants lose their leaves in the fall.

The plants make new buds at the base of each leaf's stem, but these buds do not let any water escape.

Where do animals go when it gets colder?

Many animals sleep through the winter. This is called **hibernation**. During the fall, they must find and prepare their hibernation sites. Let's do an experiment to discover which materials can help an animal keep warm so it won't freeze.

1. Fill one of the large sealable bags with dirt and leaves.

You will need

- 2 sealable plastic gallon bags
- Leaves and dirt
- 1 packet of gelatin
- Measuring cup
- 10 ounces of warm water
- 1-quart bowl
- Spoon
- 2 sealable plastic snack bags
- 8-quart or larger mixing bowl filled with ice and cold water
- Timer or clock

2. Add the packet of gelatin to 10 oz. of warm water in a 1-quart bowl. Stir with the spoon until it has dissolved.

3. Pour 2 ounces of the liquid gelatin into each small plastic snack bag. Seal them closed.

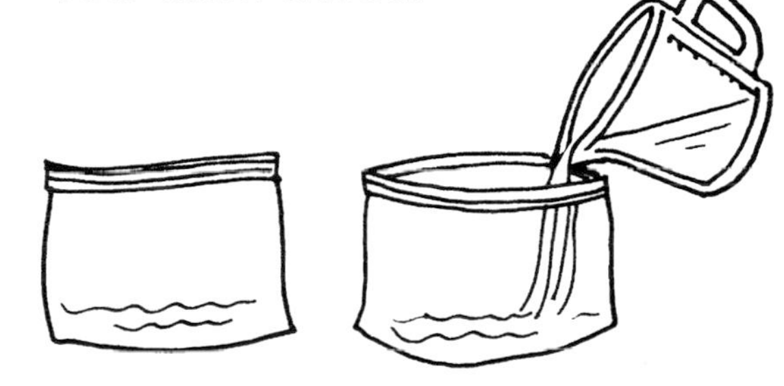

4. Place one small bag of gelatin in the middle of the leaf bag, and place the second small bag of gelatin into the empty gallon bag.

5. Place the large bags into the bowl of ice-cold water. Check the bags of gelatin after 6 minutes.

The gelatin in the empty bag is your **control** and it should harden. If the gelatin in the bag with leaves and dirt is still liquid, then that **environment** could work as a hibernating site. You can try other combinations of material in the large bag to see what works the best to keep the gelatin warm.

Bags containing gelatin that remain liquid are in environments that could work as hibernating sites. Those containing gelatin that has solidified are in environments that would not keep the animal warm enough to survive through the winter.

How do fish survive when the pond freezes?

Ice freezes downward from the top of the pond.
This keeps the water below from freezing.
Let's do an experiment to test this.

You will need

- Clear plastic cup
- Water
- Freezer
- Timer or clock

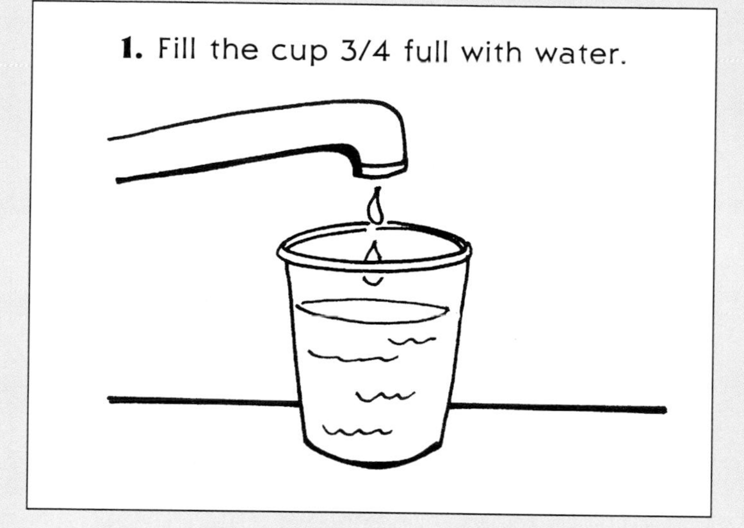

1. Fill the cup 3/4 full with water.

2. Place it in the freezer.

3. Check it every 15 minutes.

The water will freeze from the outside, going inward. A thin layer will first form on the surface. As the ice forms, it **insulates** the water inside, making it harder to freeze.

Why do we have to wear coats and gloves on cold days?

BUS
STOP

When it is cold, heat flows out of our bodies.
Materials like cloth (which are called **insulators)** make the heat flow out more slowly.
Other materials like metal (which are called **conductors**) let the heat out faster.

You will need

- A variety of objects to test: glass, plate, paper, cotton, feather, metal, aluminum foil, wood, leather, plastic
- A table indoors
- Timer or clock

1. Set all of the objects you will test on a table indoors. Space them slightly apart from one another.

2. Leave the objects there for at least 10 minutes so that they all will be at room temperature.

3. Select one object at a time and press your hand against it. Does each feel cool or warm?

When we wear coats made of cloth, the cloth helps the heat stay with our bodies and keep us warm. If our coats were made of metal, the metal would carry our body heat away and we would feel cold.

I wonder how animals
keep warm outside?

Animals have protective coverings and oils that keep water away from their skin. This keeps them dry and warm. Let's do an experiment to test this out.

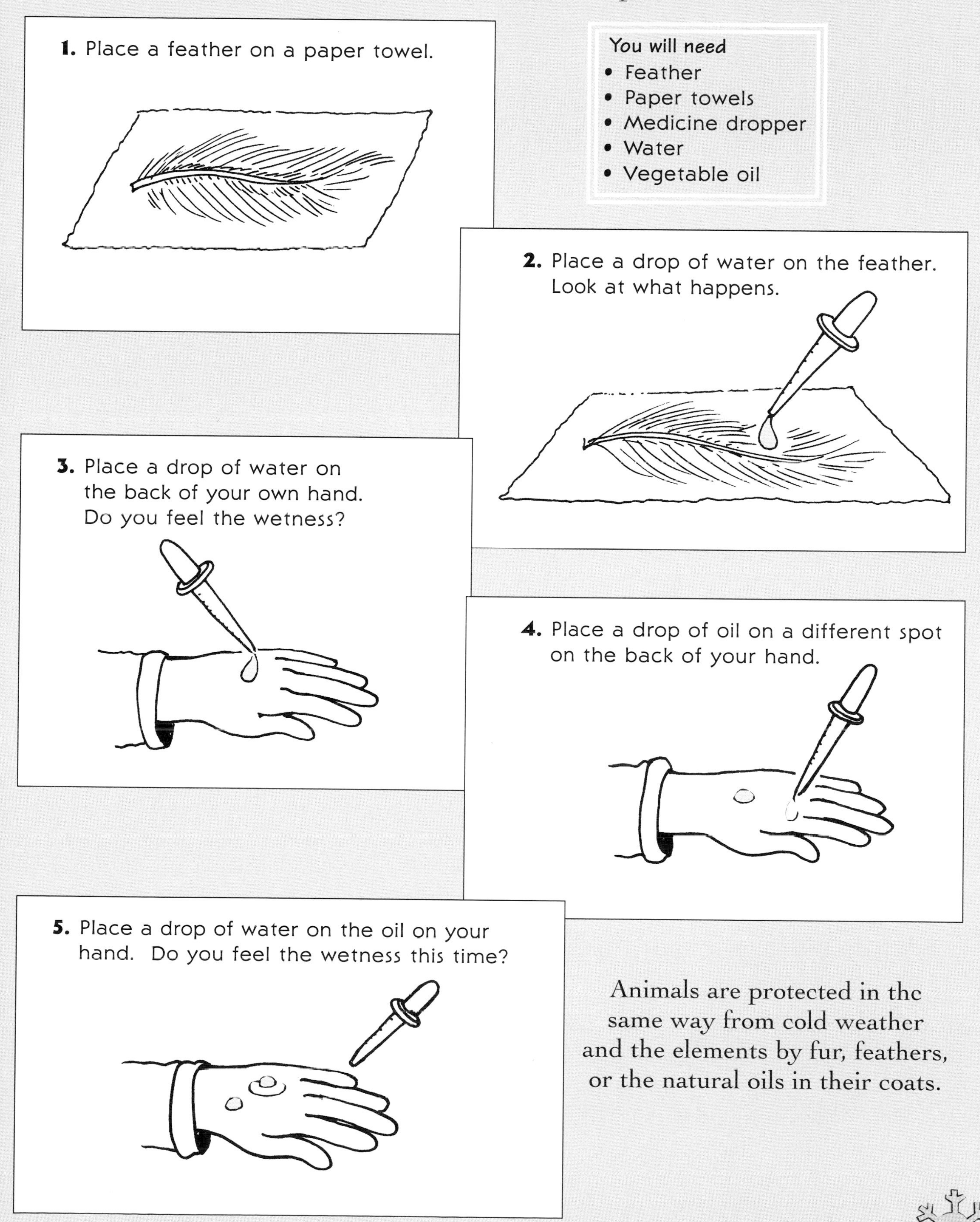

Animals are protected in the same way from cold weather and the elements by fur, feathers, or the natural oils in their coats.

How does our house stay warm?

Just as **insulation** (such as a coat and gloves) keep our bodies warm, insulators are also used to keep our houses warm. However, a house can't put on a coat and gloves! Let's do an experiment to find out what can work for a house.

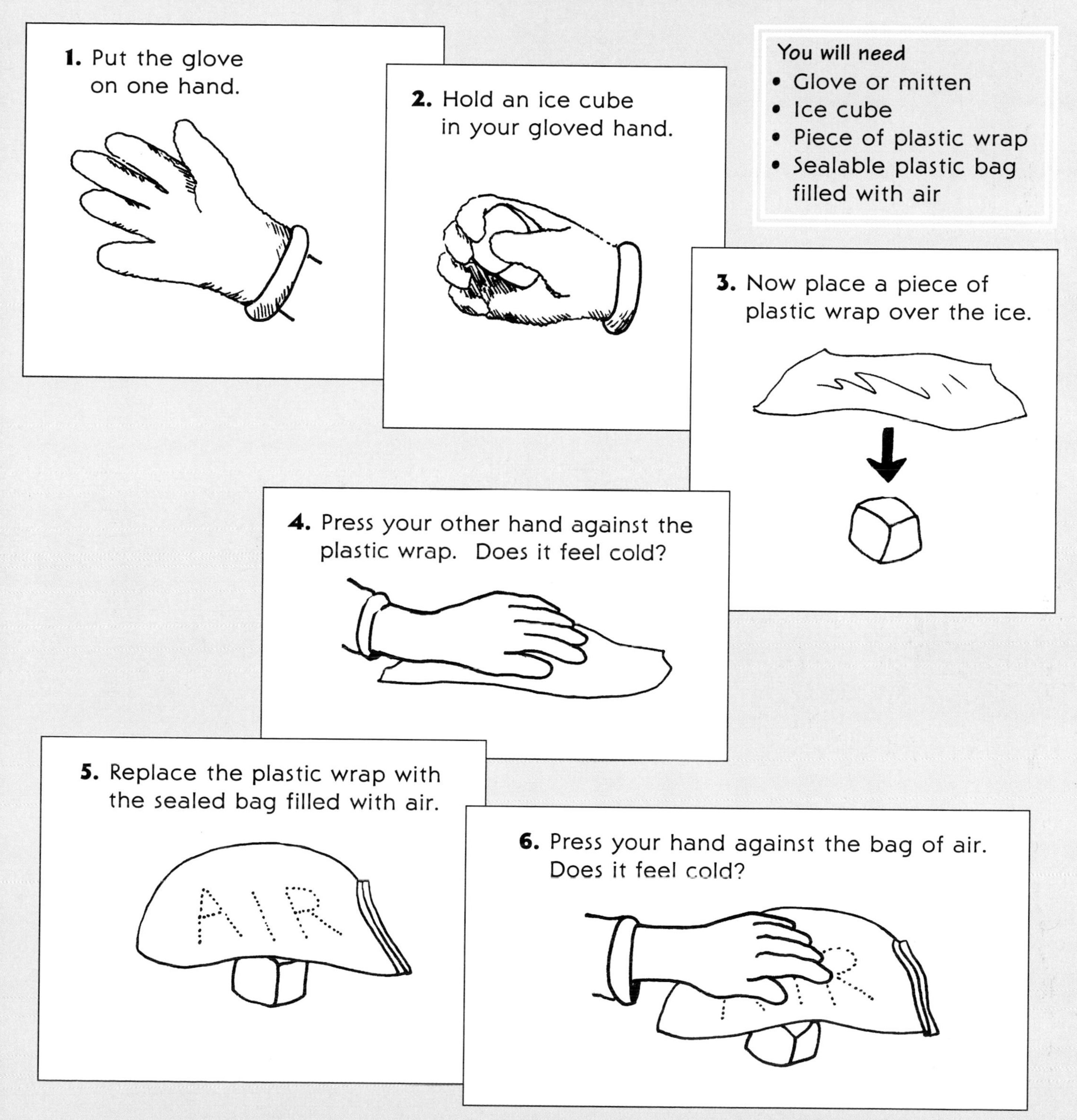

Materials that felt cold would not work to insulate your house.
The plastic wrap and plastic bag are both made of plastic, but they felt very different.
Trapped air, which was inside the bag, is one of the best insulators.
That is why storm windows and storm doors keep houses warm. They trap air!

Why do they spread salt
on the roads in winter?
ROCK
SALT

Melting snow can freeze, creating a layer of ice.
Slippery roads, driveways, and sidewalks are very dangerous.
Salt can help prevent this. Let's test this for ourselves.

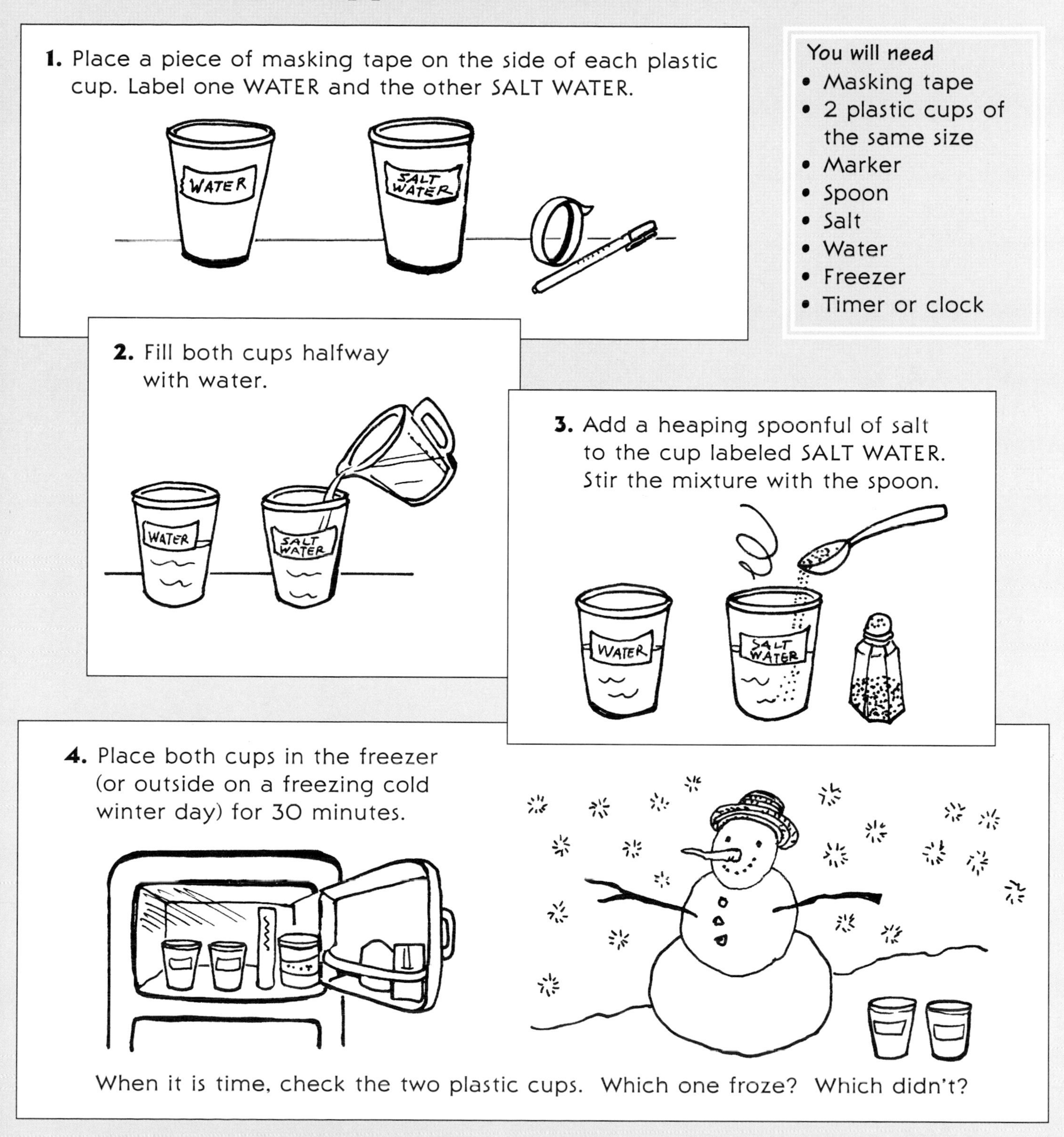

Water freezes at 32° **Fahrenheit**, or 0° **Celsius**. It has to be colder than this for salt water to freeze because the salt **crystals** get in the way of water **molecules** joining one another to make ice. Salt is spread on roads when it is cold to make it harder for the roads to become icy and slippery!

How do meteorologists know how much snow has fallen?

FORECAST

25

32

40

Meteorologists measure **precipitation** (rain or snow) by catching what falls from the sky. You can do this too, with some help from an adult.

You will need
- Ruler
- A clear 2-liter soda bottle with the top cut off by an adult
- Permanent marker
- Glue
- Old plate or block of wood
- Outdoor chair or wooden box

1. Use the ruler to measure off a series of one-inch lines on the soda bottle. Use the marker to draw the lines and write the numbers on the lines. Write big numbers so you can read them from across the room.

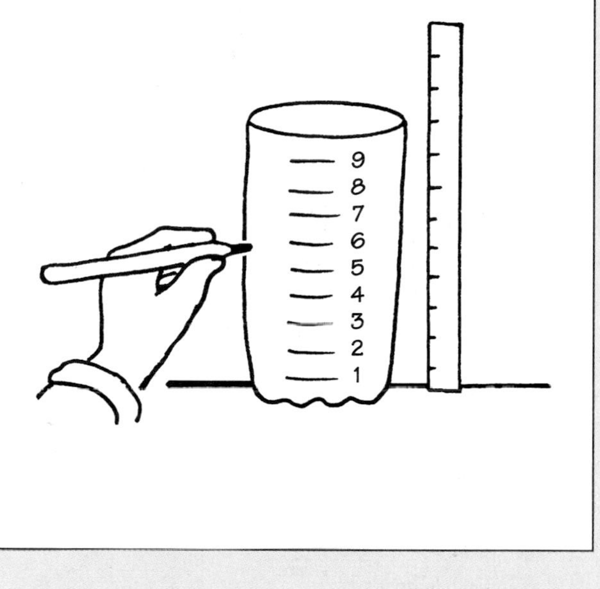

2. Glue the bottom of the bottle to a plate, or to the middle of the piece of wood. This is your snow catcher.

3. After the glue has dried, place your snow catcher outside, on an outdoor chair or wooden box, in a place where no one will step on it and where there is plenty of open sky above it.

4. When it snows or rains, measure the precipitation by seeing which line the snow or rain in the container reaches.

Empty the snow catcher between snowfalls so that you can get new readings for each one.

My birthday is here again!
One year went by fast,
watching all the seasons,
but why does my birthday always
come on the first day
of spring?
March

The Earth takes one year or 365 days to go around the sun. Each and every day the Earth is in a different position in its **orbit** around the sun. This next experiment will help explain why the same seasons follow one another in the same order every year.

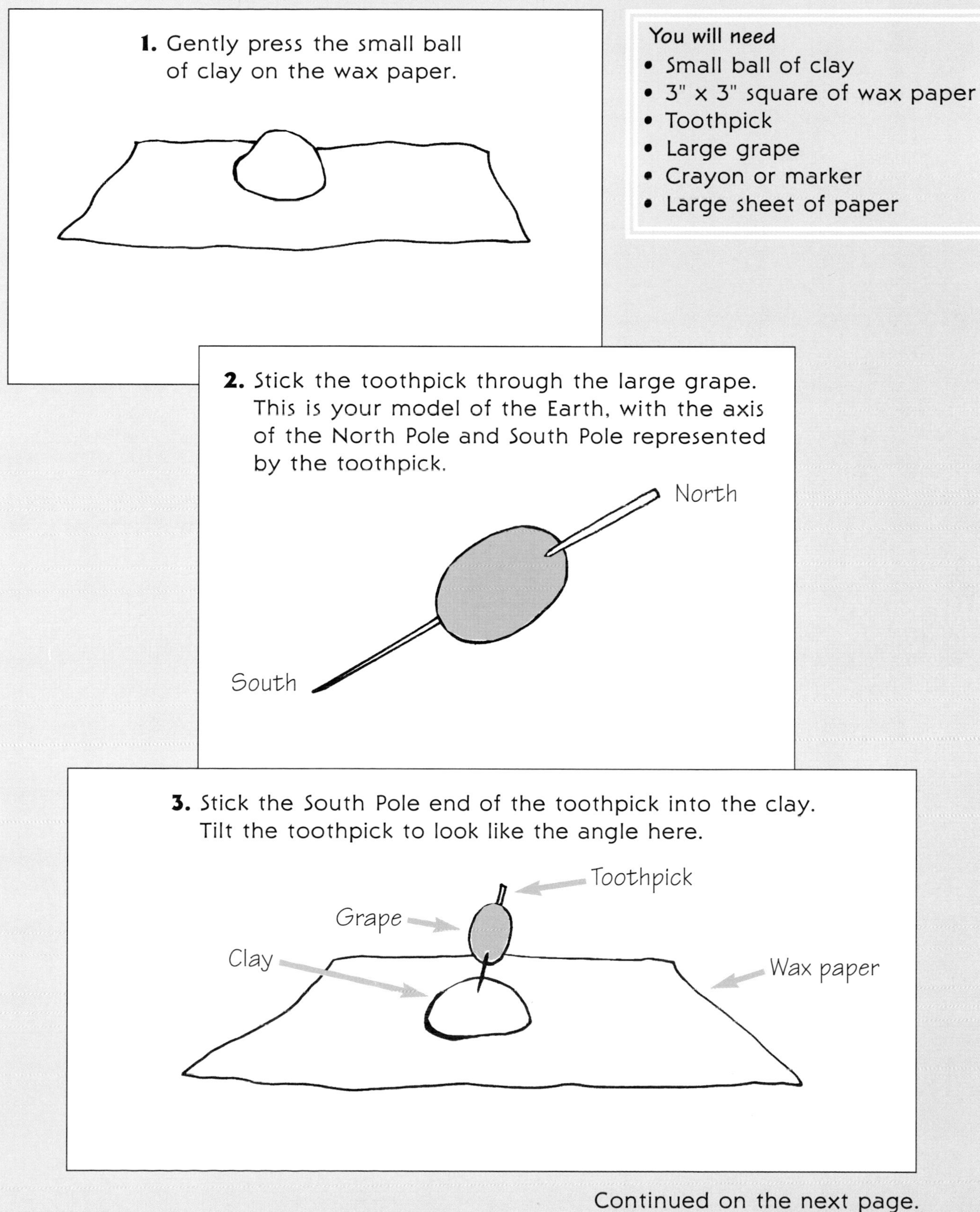

Continued on the next page.

4. Use the crayon or marker to copy this picture onto your large sheet of paper. The sun is in the middle, the Earth is in four positions around the sun, and the arrows show the path of the Earth's orbit.

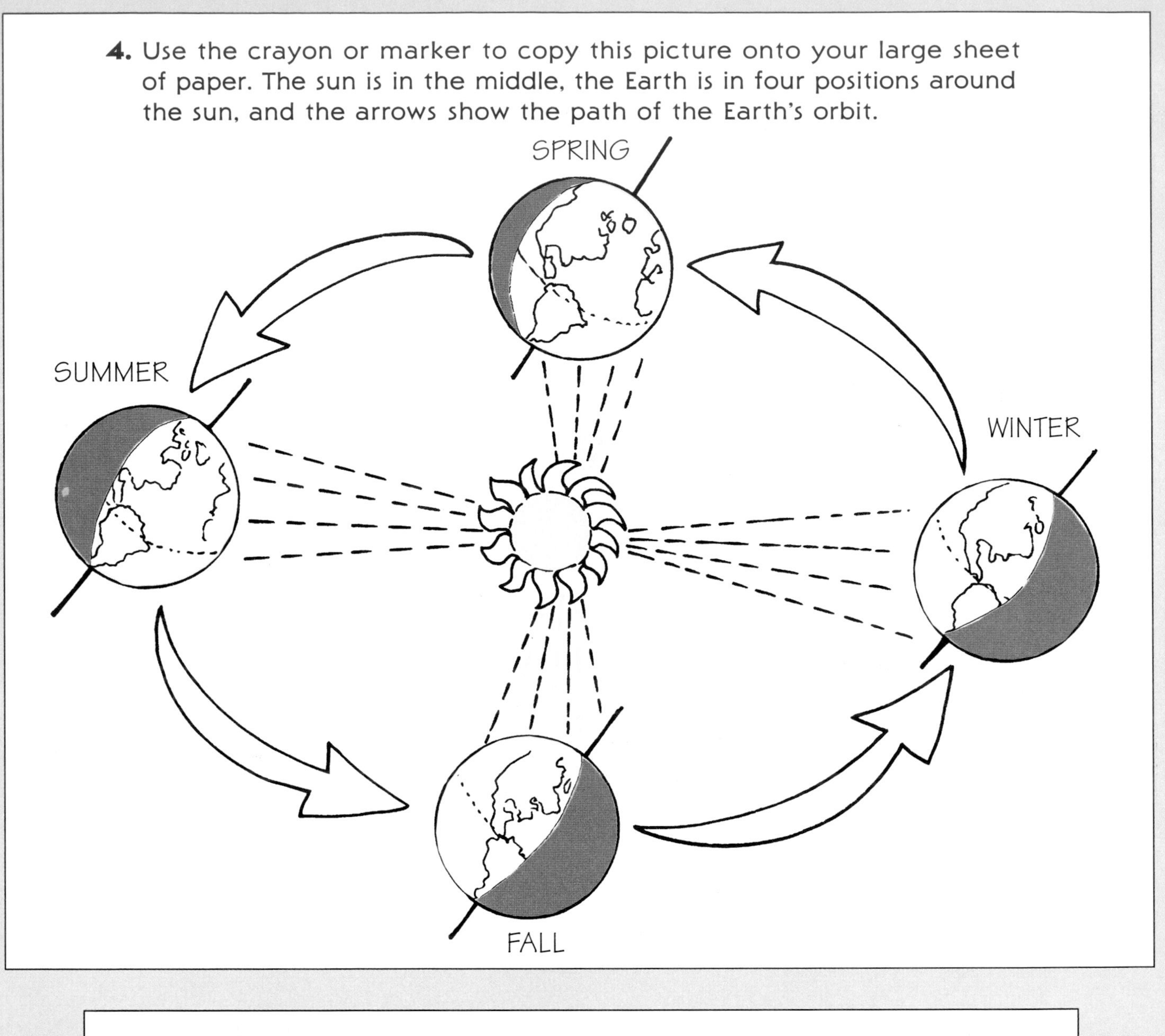

5. Place your model of the Earth, with the wax paper, on top of the Earth position that is labeled SUMMER. The North Pole should be pointed toward the sun.

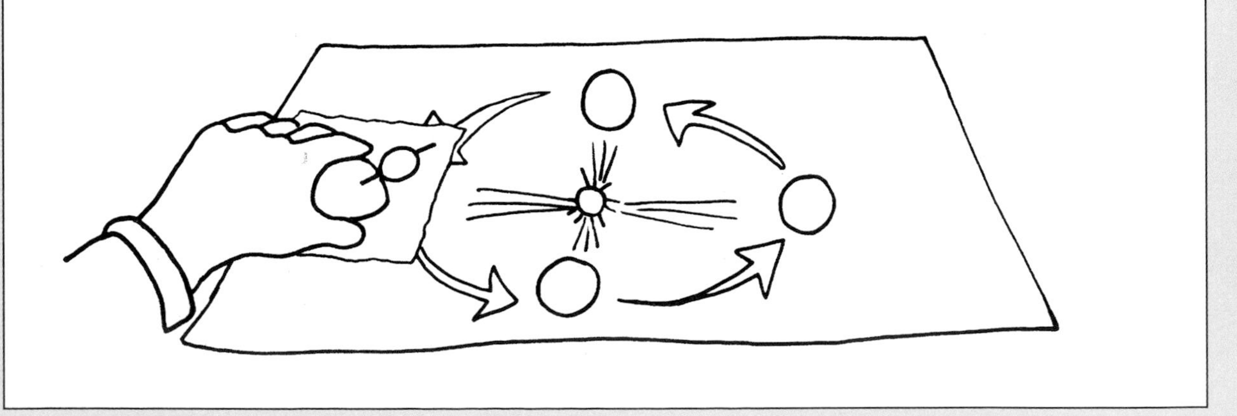

Continued on the next page.

6. Holding the wax paper, slide the Earth model around the paper, following the direction of the arrows. Keep your wrist and hand in the same position on the wax paper as you move your model in an orbit around the sun.

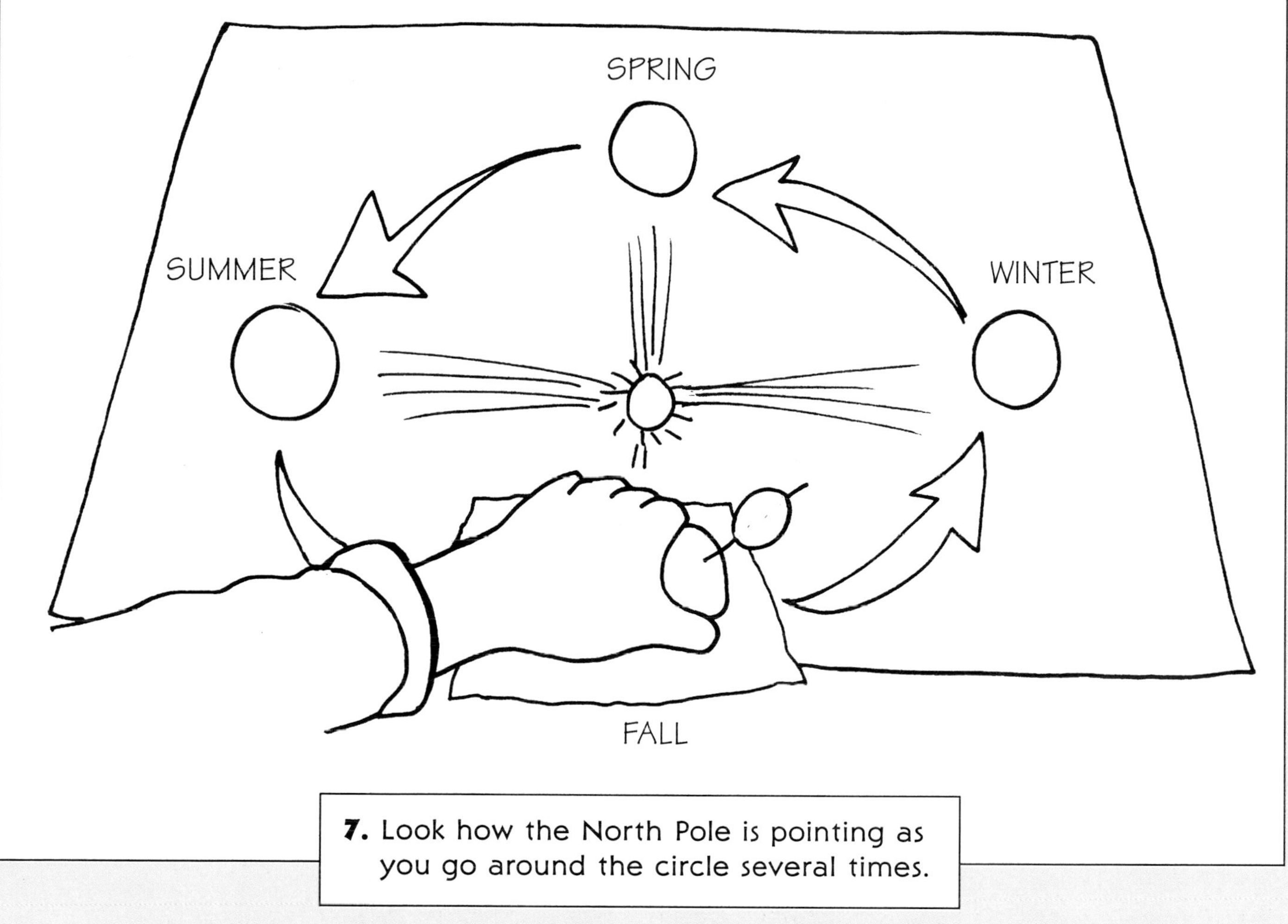

7. Look how the North Pole is pointing as you go around the circle several times.

Your hand couldn't twist around, so the North Pole stayed pointed in the same direction. This is what happens when the Earth orbits around the sun.

When the North Pole is pointed toward the sun, it is warmer in the northern half of the world and the season is summer. When the North Pole is pointed away from the sun, it is colder in the northern half of the world and the season is winter.

Spring and fall are the seasons when the North Pole is not facing toward or away from the sun. At these times the side of the Earth is toward the sun.

Although this is the last experiment in this book, keep observing what is happening around you. Test things out by doing your own experiments. Maybe you'll even discover something new!

Glossary

albumen — the white, or clear part of an egg

Celsius — a thermometer scale with 0° as freezing water and 100° as boiling water.

chromatogram — a picture produced by doing chromatography

chromatography — a method scientists use to separate materials that are different

conductors — materials that transfer heat

control — an untreated sample used for comparisons in experiments

crystals — solid substances with their molecules in repeating patterns

dispersal — the process of scattering something in small parts

dissect — to cut apart, piece by piece, for inspection

environment — the surroundings and conditions that affect the survival of a plant or animal

Fahrenheit — a thermometer scale with 32° as freezing and 212° as boiling water

germination — seeds beginning the process of sprouting

hibernation — animals spending the winter in a restful or sleeping state

insulation — something that prevents the escape or transfer of heat

insulators — materials that stop the transfer of heat

microscope — an instrument used to look at very small things

molecules — smallest individual particles of a substance

nectar — the sweet liquid in many flowers that bees use to make honey

orbit — the path the Earth travels around the sun once a year

petals — the leaf-like parts of a flower

pigments — coloring matter in animals, plants, or paints

pistil — the part of a flower that produces seeds

pollen — the yellow, powdery particles in flowers that are needed to germinate seeds

precipitation — water coming from the sky (rain, snow, sleet, hail, etc.)

solar — dealing with the sun

stamen — the part of a flower that produces pollen

stigma — the upper tip of the pistil of a flower that receives pollen

thermometer — an instrument used to measure how hot or cold it is

vapor — liquid molecules floating in a gas

yolk — the yellow part of an egg